Daring to Dig

Paleontological Research Institution Special Publication No. 54

ISBN 978-0-87710-521-3

Library of Congress Control Number 2017952162

Illustrations by Alana McGillis
Book design by Beth Stricker

Printed and bound in the USA
Second printing December 2018

Published by the Paleontological Research Institution
1259 Trumansburg Road
Ithaca, NY 14850, USA
Visit www.priweb.org
and www.daringtodig.org

Daring to Dig

Adventures of Women in American Paleontology

by Beth Stricker
illustrated by Alana McGillis

PALEONTOLOGICAL
RESEARCH INSTITUTION

FOSSILS!

the remains of past living things usually preserved in rock

plants

animals

tracks

Paleontologists are scientists who uncover and study these clues to learn about life long ago. Thomas Jefferson, third president of the United States, was one of America's first paleontologists. He studied fossils, from the giant ground sloth to the enormous mastodon. Since then, paleontologists have explored Earth's mysteries.

Most paleontologists have been men. There were many reasons for this, but the biggest was that many men did not think women should have jobs, go to college, or become scientists. Often they discouraged women from chasing their dreams.

A few women broke these rules and pursued their passion. With support from their families, some went to college. If they were offered a job afterwards, they earned less than men. Over time, several bold, smart women created their own paths to paleontology and became role models for others.

Despite facing discrimination, women have shaped our understanding of the history of life on Earth. These stories highlight some of the biggest adventures of their careers. So, if you dare, let's dig in!

Annie Alexander

1867 to 1950

In 1903, adventurer Annie Alexander was on an expedition in Shasta County, California with a group of friends. The team uncovered the bones of a prehistoric creature with paddle-like arms and legs.

With money from her father's successful business and her own wise investments, Annie could afford to travel the world and donate specimens and funds to the University of California. She and Louise donated more than 20,000 fossils and modern plants and animals to the University's collections.

Thalattosaurus alexandrae
Tha-la-toe-SORE-us a-lex-AN-dree

Mignon Talbot

One morning, Mignon was out for a stroll with her sister, Ellen, near Holyoke, Massachusetts. They passed a hill of gravel and Mignon noticed a large piece of sandstone streaked with white. She stopped to take a closer look and discovered...

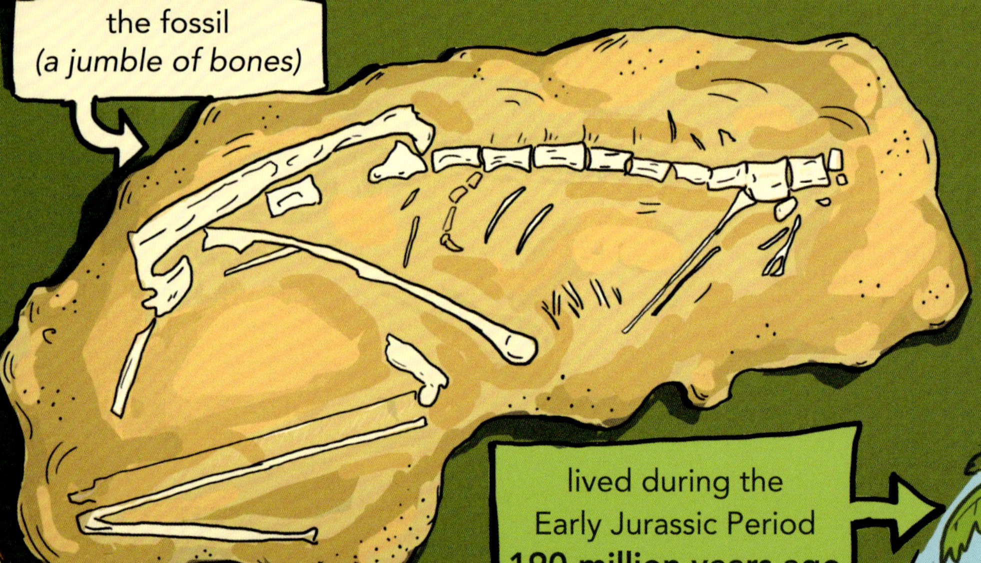

What did *Podokesaurus* look like? Did it have feathers like a bird? Rough skin like a lizard? Or maybe something else entirely!
Mignon pictured the dinosaur with its long tail dragging on the ground…
But scientists now think it would have been held out for balance.
I think I'll call you *Podokesaurus holyokensis*, meaning 'swift-footed lizard of Holyoke'. I bet you were a fast little thing!
PROTO FEATHERS
The first woman to find and describe a dinosaur fossil!
ROCK HAMMER
Podokesaurus holyokensis
POH-doh-kuh-SAWR-us HOL-ee-o-KEN-sis
The fossil was incomplete and most of the skull was missing. But Mignon examined the evidence and published a paper about her discovery in 1911. The dinosaur would have walked on two feet, and was the same height as a chicken, though much longer.
It was a **carnivore,** or meat-eater, that probably hunted in small groups.

Carlotta Maury

1874 to 1938

Carlotta loved traveling. In 1916, she headed to the Dominican Republic to collect a variety of fossils and learn more about the age of the rock around them. Many of these fossils had not been given a name yet. She and her team arrived in the midst of a revolution. But that didn't stop them.

Carlotta wanted to know the **stratigraphic sequence**: how old are the different layers of rock?
The fossils tell us!

Karl and Axel were recent graduates from Cornell University. Professor Harris sent them along to help and protect Carlotta.

By studying the fossils she had found, Carlotta could tell how old the different layers of rock were in the Dominican Republic.

Most of the fossils she collected in the Cibao Valley formed between 4 and 7 million years ago.

She collected fossils of invertebrate animals (no backbone!):

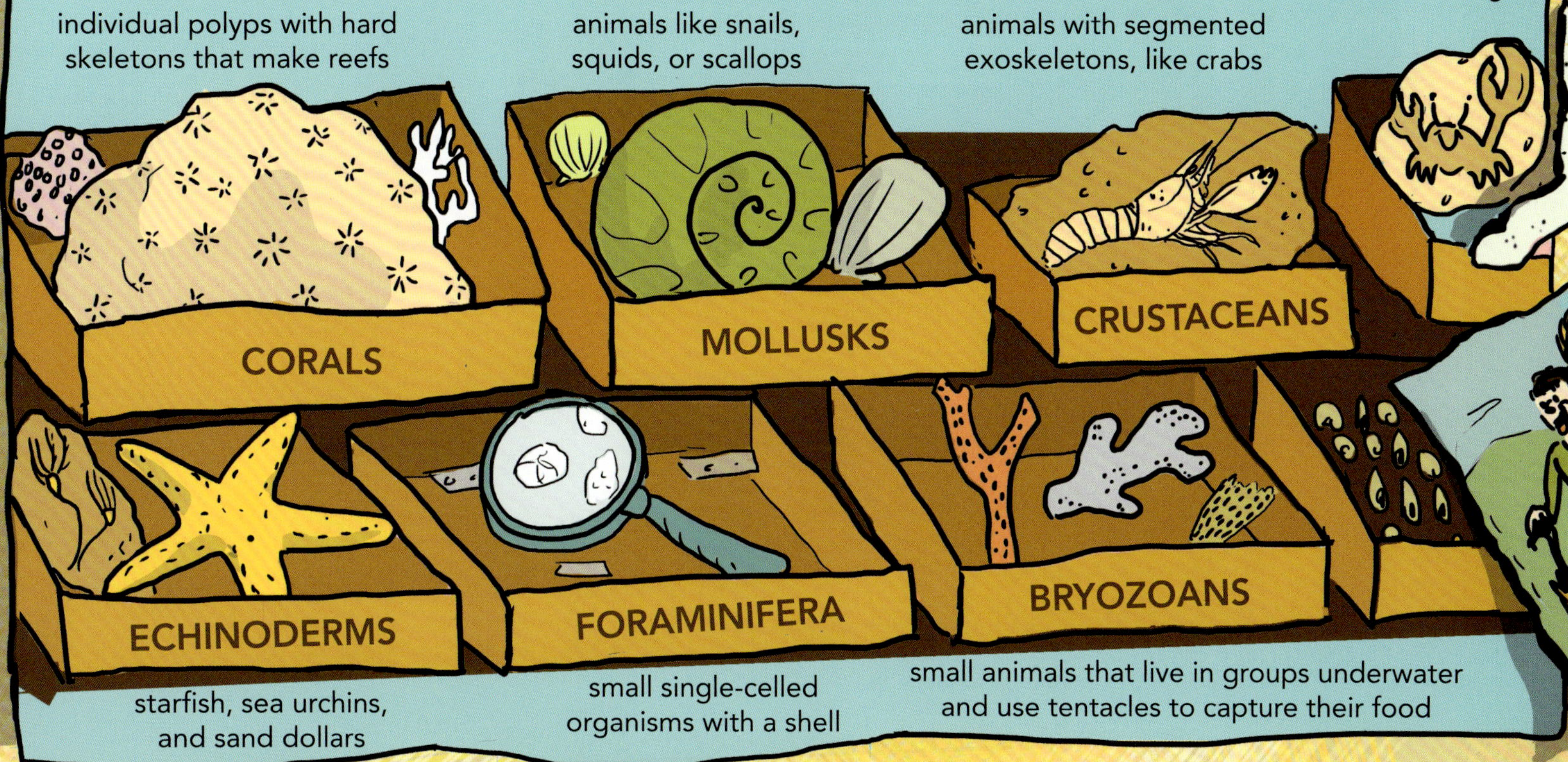

Normally she wore dresses, but these “bloomers” – designed by fellow paleontologist **Winifred Goldring** – were perfect for adventures.

Carlotta had earned her PhD at Cornell University, and returned often to conduct research.

That summer, Carlotta found out that another group of scientists planned to publish about these new fossils. Whoever finished first would get the credit. The race was on! For months she worked, writing 300 pages with 40 illustrations. She and her professor, Gilbert Harris, published her writings just before the other team.

Julia Gardner

1882 to 1960

TOP SECRET

In 1944, the world was at war. Allied forces pushed the Germans east from France. American troops held off Japanese armies in the Pacific. Meanwhile, the Japanese floated hundreds of dangerous balloons over western North America. Each one carried a bomb. Julia had to figure out where they were coming from.

DANGER

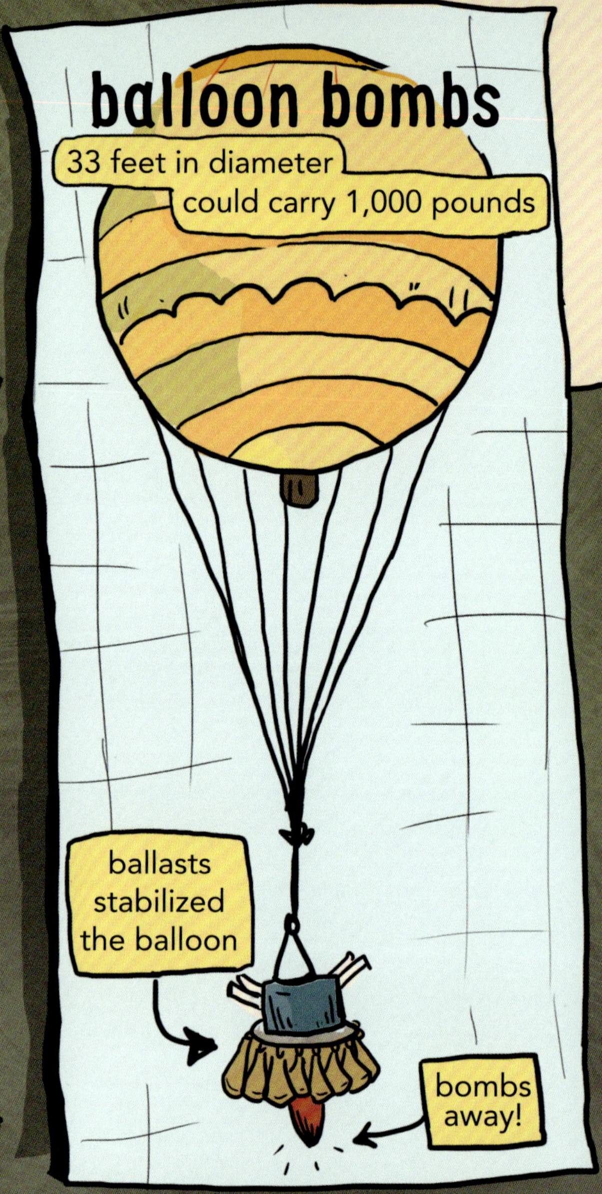

Julia joined the United States Geological Survey's **Military Geology Unit**
The MGU helped war efforts by making maps and identifying water sources and airplane landing sites.

For most of her time at the USGS, Julia studied the **Alum Bluff Group in Florida** and mapped the geology of the **West Pacific Islands**.

the balloon flew for three days before the fuse lit automatically
ballasts full of sand
a 33 pound anti-personnel fragmentation bomb
PLONK
JAPAN
• = Balloon landing
USA
Hmm, all of the beaches along this coastline have bits of coral in their sand, except these three...
What was the sand made of?
MOLLUSKS
animals like snails, slugs, or scallops
FORAMINIFERA
single-celled organisms with shells
CORALS
individual polyps with hard skeletons that make reefs
Julia and her team in the MGU collected sandbags from unexploded bombs. She examined the sand used as ballast and observed that there were no bits of coral in it. That meant the balloons were being launched from a beach northeast of Sendai, Japan.
US ARMY
DECLASSIFIED

Winifred Goldring

1888 to 1971

In 1924, Winifred was teaching classes at Wellesley College when she was invited to come work for the New York State Museum. After two years as a scientific expert, Winifred undertook a massive project: a book about the state's Devonian crinoids.

CRINOIDS
echinoderms, related to starfish, sand dollars, and sea urchins

In the Devonian, about 380 million years ago, sea levels were high worldwide. Crinoids lived in shallow seas that covered much of the land.

Over the years, scientists had collected notes, drawings, and specimens, but none had completed the project. Winifred continued to collect and organize information. After seven years, she published a detailed book on the topic.

Although she completed many other major projects at the Museum, Winifred was always known as an international expert on Devonian crinoids.

1895 to 1972

Esther Applin

At the 1921 meeting of the Paleontological Society, Esther presented revolutionary research. She suggested a new way of finding oil based on the age of tiny fossils found nearby. Older scientists dismissed the notion, but not for long.

These three women led the launch of a new field of work: applied micropaleontology.

MICROPALEONTOLOGY
the study of fossils too small to be seen in detail without a microscope

Using microfossils, they described the layers of rock and oil-bearing zones in the Gulf Coast.

Esther wanted to use stratigraphic correlation to locate new sources of oil.

STRATIGRAPHIC CORRELATION
Comparing one area of rock to others to find out how old it is. Layers with the same fossils are the same age.

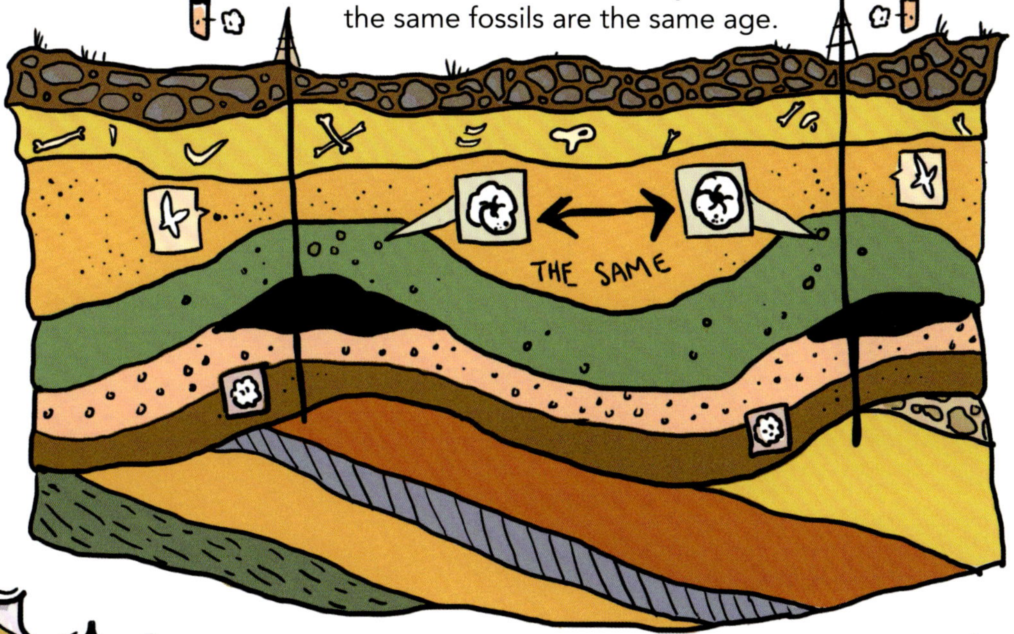

How is oil made?
Millions of years ago, algae died and were buried.

Over time, and with lots of heat and pressure, the algae transformed into oil.

Once taken out of the ground, petroleum powered ships, trucks, tanks, and much more.

Esther worked closely with her friends, Alva and Hedwig. They collected information to show that microfossils were useful in stratigraphy and correlation. In 1925, they published a paper that proved this new method *could* be done and *had* been done.

Katherine Palmer

As Director of the Paleontological Research Institution, Katherine had bold ideas about finding, organizing, and studying invertebrate fossils. By the late 1950s, the Institution's collections and activities had outgrown their home near Cornell University. Katherine made ambitious plans for the future.

1895 to 1982

Katherine lived in Ithaca for most of her life, studying fossil clams and snails that lived 34 million to 66 million years ago.

INVERTEBRATES animals without a backbone

Professor Gilbert Harris, who mentored Katherine at Cornell University, founded PRI in Ithaca, New York.

Katherine collected fossils in the US, the Gulf of Mexico, West Indies, and New Zealand

SCIENTISTS USE COLLECTIONS...
...to notice change over time

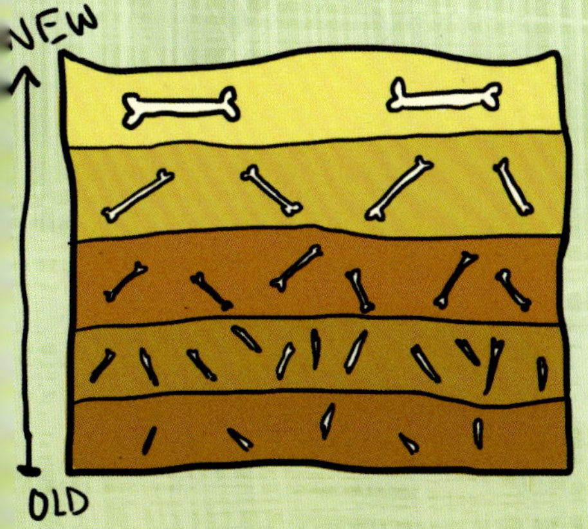

...to identify known species and newly discovered ones

...to map out relationships between different species

Katherine spent less time on scientific research after becoming the director, but her new work helped others make discoveries. She improved the library, the research lab, storage of the fossils, and more. She also became an enthusiastic educator for the people of Ithaca.

1897
to 1967

Tilly Edinger

In 1940, Tilly stepped off a ship onto the busy streets of New York City. As a Jewish woman, the past few years had been terrifying for her. She had been an unpaid curator in a German museum when strict Nazi laws forced her into hiding. Tilly found temporary safety in London until sailing for the United States on the SS Britannic.

Threatened by Nazi rule, Tilly worked quietly at the museum until 1938 when she was in immediate danger. And so she fled.

Tilly sorted and organized fossils, and also studied the shape and size of fossil skulls.

She had worked at the **Senckenberg Museum** in Frankfurt, Germany

Paleontologist George Gaylord Simpson supported Tilly's immigration to the US.

After sailing to New York, Tilly moved to Boston, where she continued the work she had started in Germany.

BOSTON
The brain was not fossilized... So how do I study it?
Tilly started losing her hearing when she was young. She used hearing aids for most of her life. This made speaking at scientific conferences or working as a teacher difficult.
She arrived at Harvard University just in time to be a founding member of the **Society of Vertebrate Paleontology,** December 1940.
By making an **endocast!**
POP!
The inside of a fossil skull has imprints of the brain that it once held.
Fill the skull cavity with plaster or rubber.
Once dry, compare the endocasts to the brains of living animals.
NAME:
BRAIN:
HORSE:
Today
Equus (modern horse)
Pliohippus
Merychippus
Mesohippus
Eohippus
so small
10 Million Years Ago
ENDOCAST!
Tilly embraced her new life in Boston. She found that American paleontological sites held the remains of extinct horses. Tilly spent ten years collecting endocasts of these brains and studying them. She wrote an important book on the evolution of the horse brain.
She focused on the horse family, called **Equidae.**
Tilly founded a new scientific field. **Paleoneurology:** the scientific study of the nervous systems and brains of extinct animals.

Mary Dawson

There is rock beneath Ellesmere Island, in the Canadian Arctic, that is 55 million years old. Mary read a report on the geology of the region and thought that there might be fossils preserved in the rock. In 1973, she and her team made the long journey to the island.

PLATE TECTONICS
the large-scale motion of Earth's crust, which is divided into separate plates, over hundreds of millions of years

Mary's fieldwork uncovered evidence that the region had been a warm, frost-free swampland 55 million years ago. Fossils showed that mammals had traveled from North America to Europe on land that no longer exists. Her work provided support for the theory of plate tectonics.

Karen Chin

Paleontologists search for fossils, clues to life long ago. Some fossils are shells, some are bones, and some are footprints. Deep in the rolling hills of northwest Montana, Karen uncovered a special treasure. Coprolites... fossilized poop!

How did she know it was petrified poop? By the distinctive burrows in and around the dung.

These coprolites are about the size of a basketball.

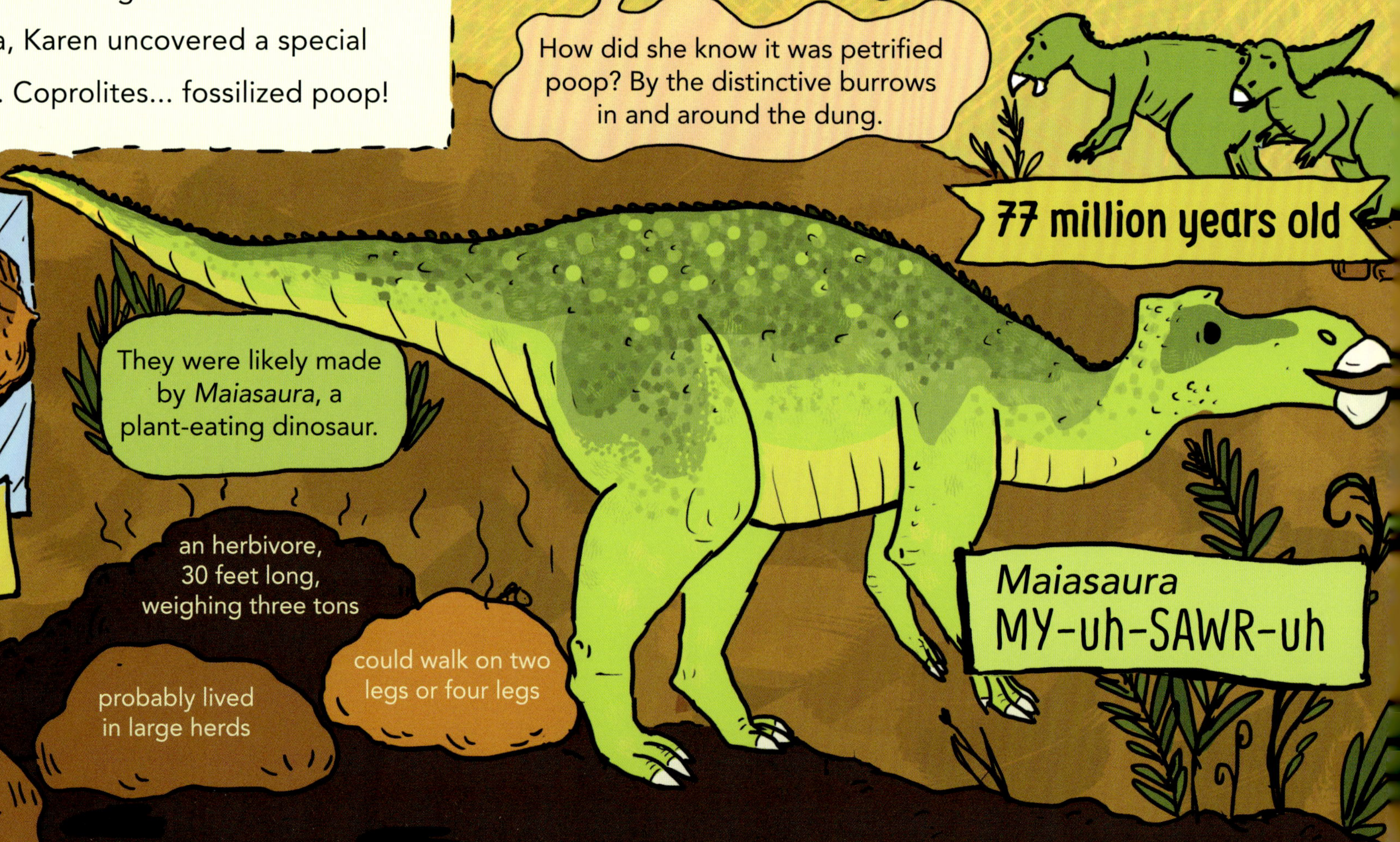

Most of the time, Karen uses a microscope to examine thin sections of fossil feces.
She studied these coprolites and found shredded wood fragments – **the *Maiasaura* ate rotted wood!**
Sections of other fossil feces might reveal: pollen grains, seeds, leaves, fish scales, snail shells, muscle tissue, teeth, plant spores, bacteria or bone fragments.
These clues help reveal which plants and animals lived at the same time as the dinosaurs, what they ate, and what the climate was like.
For every hour of fieldwork, you create hours of lab or collections work.
at the University of Colorado
CHEMICALS
x-ray
CT scan
Different techniques in the lab reveal the chemical makeup of the coprolites.
The results help us learn how the soft poop was turned into rock.
After she collects fossils in the field, Karen heads back to the lab to learn more. She sometimes uses high-tech tools to examine the coprolites and find evidence about the environment in which they were made. Clues within the coprolites tell us what life was like on Earth millions of years ago.

Helen James

Hundreds of years ago, prehistoric humans settled on islands in the Pacific Ocean, including Hawaii. These people affected their environment so quickly that a huge number of birds went extinct. What happened? Helen was determined to find out.

Once home to more than 110 native species of birds, the islands now hold fewer than 30. Fossil evidence shows that these extinct birds ranged from crows and owls to flightless waterfowl.

Flightless birds, foraging for food amongst the undergrowth, fell into lava tube caves.

Trapped within the caves, these birds died and were eventually preserved as fossils.

lava tube caves

Lava flowing from a volcano thickens and slows, creating a hard crust, while hotter lava continues to flow beneath it. When the hot lava is gone, a cave-like tube remains.

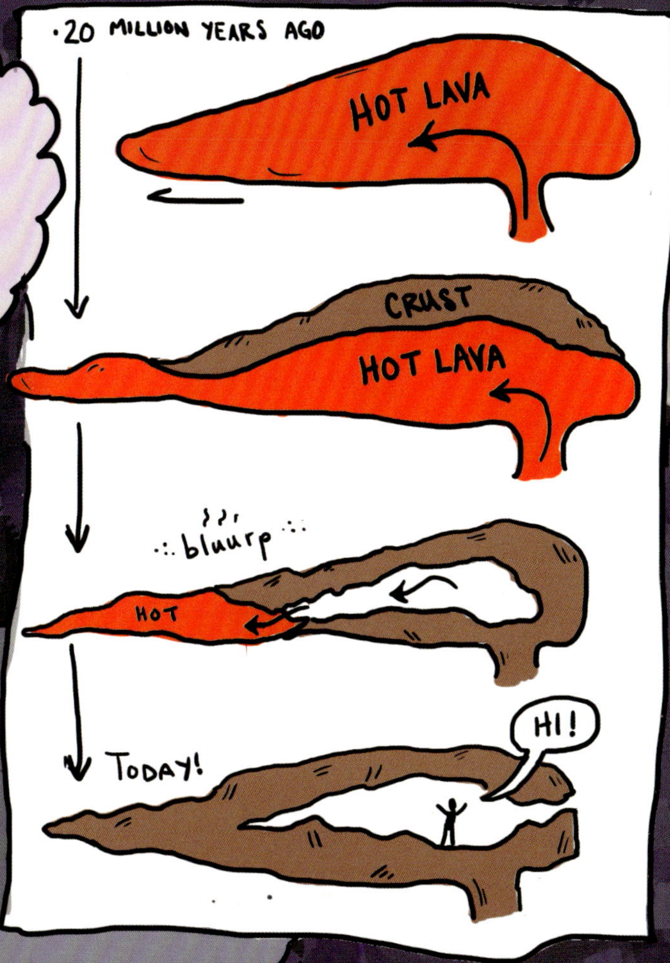

Possible causes... all associated with people

Why did these birds go extinct?

They disappeared sometime after people first appeared on the islands...

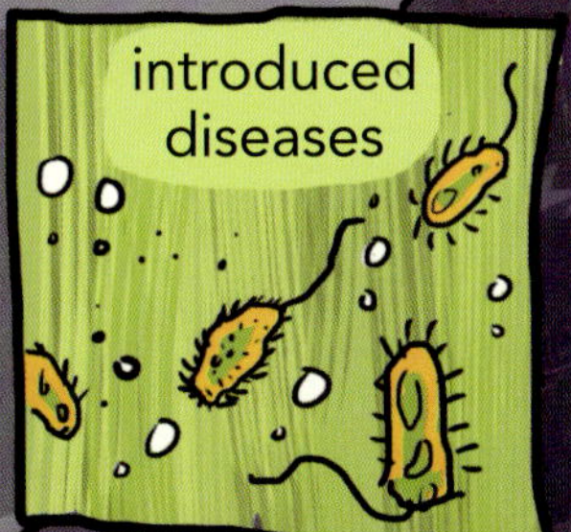

Since 1976, Helen has worked with scientist Storrs Olson to identify fossil birds. Eventually they realized that the islands held a huge number of undescribed fossils. The two made their careers documenting these species and the role humans played in the birds' extinction.

Phoebe Cohen

In the late 1970s, scientists discovered some of the oldest known fossils in the world. At the time, these tiny fossils were thought to be between 520 million to 800 million years old. Thirty years later, graduate student Phoebe took another look at these strange specimens.

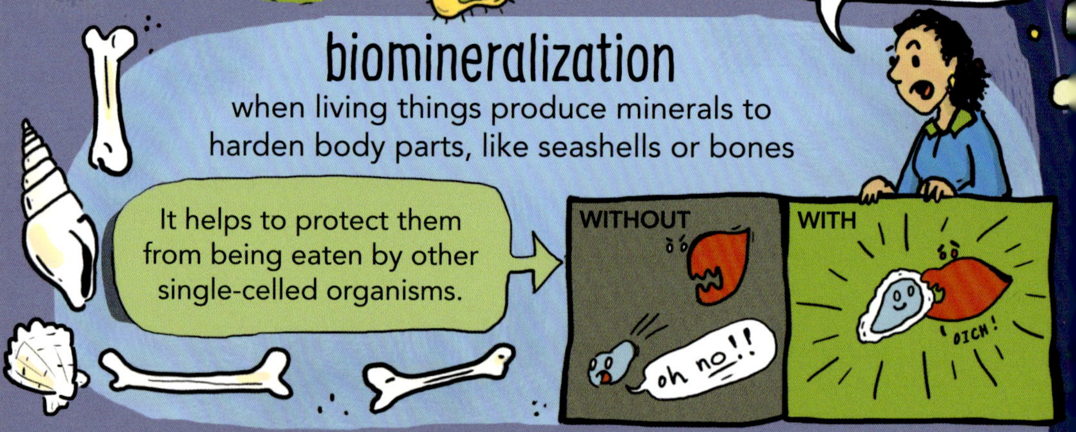

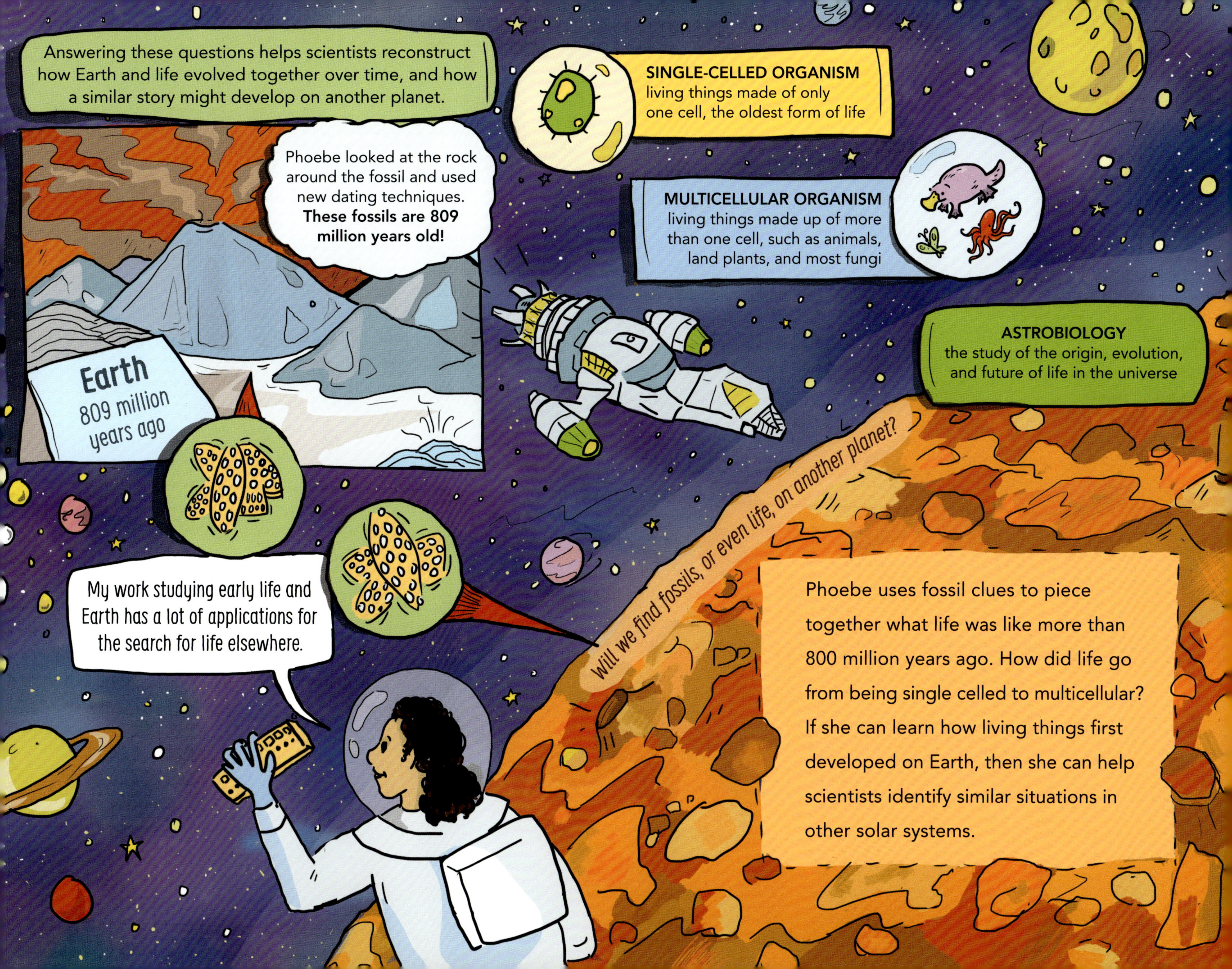

Answering these questions helps scientists reconstruct how Earth and life evolved together over time, and how a similar story might develop on another planet.
SINGLE-CELLED ORGANISM
living things made of only one cell, the oldest form of life
Phoebe looked at the rock around the fossil and used new dating techniques. These fossils are 809 million years old!
MULTICELLULAR ORGANISM
living things made up of more than one cell, such as animals, land plants, and most fungi
Earth
809 million years ago
ASTROBIOLOGY
the study of the origin, evolution, and future of life in the universe
Will we find fossils, or even life, on another planet?
My work studying early life and Earth has a lot of applications for the search for life elsewhere.
Phoebe uses fossil clues to piece together what life was like more than 800 million years ago. How did life go from being single celled to multicellular? If she can learn how living things first developed on Earth, then she can help scientists identify similar situations in other solar systems.

PHANEROZOIC
TODAY
0
500
1000
1500
2000
2500
3000
3500
4000
4500
MILLIONS OF YEARS
PHOEBE'S COMPLEX CELLS
PROTEROZOIC
ARCHEAN
LIFE!
3.8 Billion Years!
HADEAN
EARTH STARTS HERE
Karen's Maiasaura Poop
Mignon's Podokesaurus
TODAY
0
50
100
150
200
250
300
350
400
450
500
CENOZOIC
CRE-TA-CEOUS
JURASSIC
TRIASSIC
PERMIAN
CARBONIFEROUS
DEVONIAN
SILURIAN
ORDOVICIAN
CAMBRIAN
ANNIE'S THALATTOSAURUS
WINIFRED'S CRINOIDS

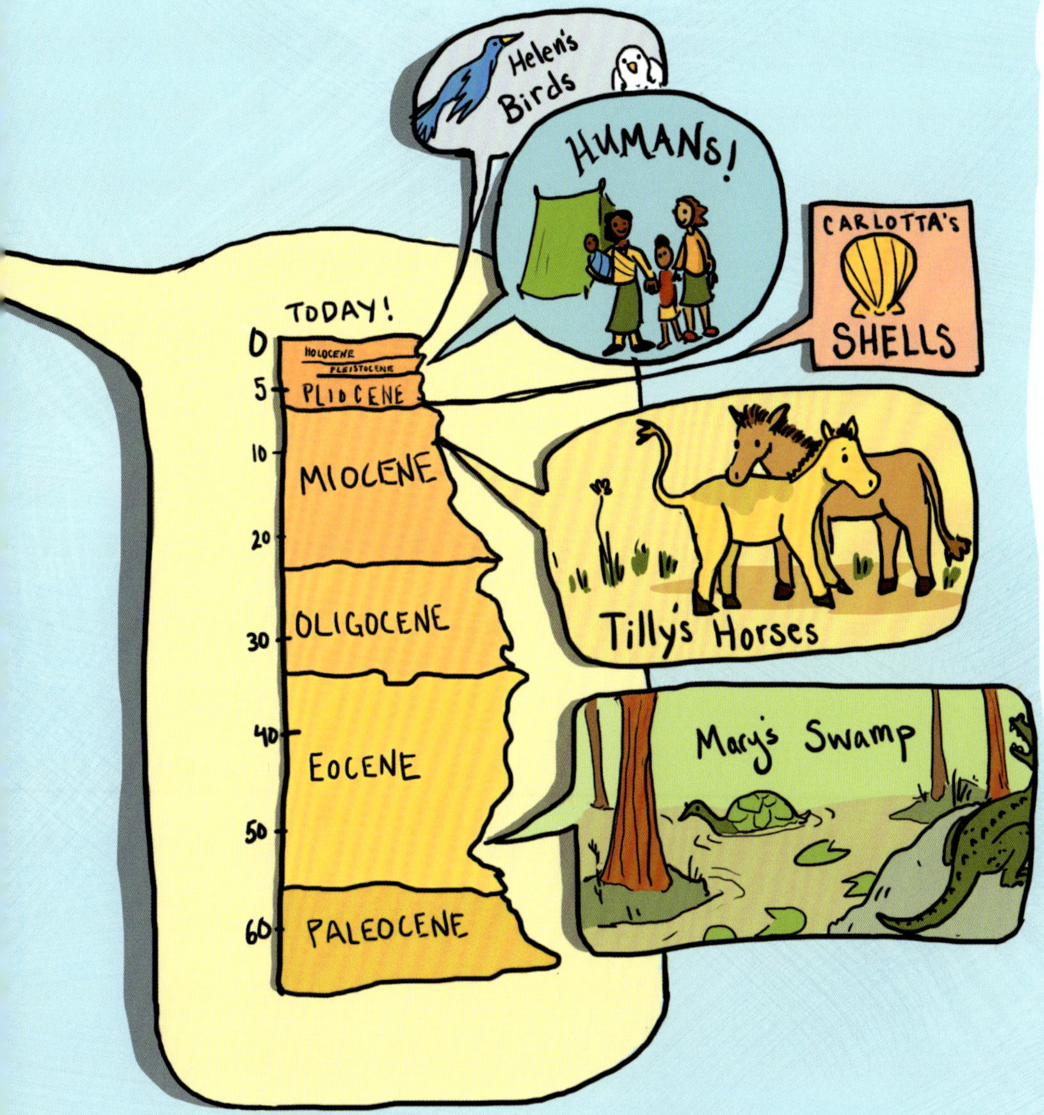

Earth is more than 4.5 billion years old, and for the last 4 billion of those years it supported life. Our planet has a vast history, especially compared to the brief amount of time humans have lived here. It can be tough to visualize just how long ago different plants and animals lived. This "map" of geologic time helps us picture the immensely long history of life.

What we know about Earth's history has changed with each new discovery. The people who make these discoveries have changed too. Despite discrimination, the field has become more diverse. As more women—of all backgrounds—dare to dig, their discoveries will bring us closer to piecing together the full history of life on Earth.

Annie Montague Alexander (1867-1950) was born in Hawaii and traveled all over the world with her family. Her interest in paleontology began when she sat in on a course at the University of California-Berkeley and she quickly became an avid fossil collector. Later she founded the University of California Museum of Paleontology and the Museum of Vertebrate Zoology.

Photo courtesy of the University of California Museum of Paleontology

Mignon Talbot (1869-1950) was born in Iowa. She received her undergraduate degree at The Ohio State University and her Ph.D. at Yale University. After graduating, she was hired as a professor of geology at Mount Holyoke College and became chair of the geology department in 1908. By the time she retired in 1935, she had built an impressive specimen collection at the college.

Mount Holyoke College Archives and Special Collections

Carlotta Maury (1874-1938) was born in Hastings, New York. She received her Ph.D. in geology from Cornell University in 1902 and focused on fossils in South and Central America. Because of her expertise in mollusk fossils, Maury eventually became one of the first women to be employed by an oil company (Royal Dutch Shell Petroleum Company).

Archives of the Paleontological Research Institution

Julia Gardner (1882-1960) was born in Chamberlain, South Dakota. In 1911, she completed her geology Ph.D. at Johns Hopkins University. She spent much of her career researching Cenozoic mollusks as a scientist at the United States Geological Survey. She served as president of the Paleontological Society and vice president of the Geological Society of America.

© Bachrach

Winifred Goldring (1888-1971) was born near Albany, New York. She completed undergraduate and graduate work at Wellesley College and received additional training at Columbia University and Johns Hopkins. At age 26 she became a staff paleontologist at the New York State Museum, where she spent the next four decades. Goldring eventually became the State Paleontologist of New York.

Photographer unknown

Esther Applin (1895-1972) grew up on Alcatraz Island, where her father was directing the construction of the infamous prison. She studied paleontology at the University of California-Berkeley before starting her successful career alongside colleagues Hedwig Kniker and Alva Ellisor. She worked for oil companies, was a college professor, and researched the geology of the southeastern United States as a member of the U. S. Geological Survey.

Image from Journal of Paleontology, v. 47(2),
© SEPM Society for Sedimentary Geology 1973, used with permission

Katherine Palmer (1895-1982) was born in Oakville, Washington. She grew up interested in natural history and was the only member of her high school class to go on to college, in her case the University of Washington. Palmer received her Ph.D. at Cornell University in 1925, before becoming the second director of the Paleontological Research Institution. She was the first woman to be awarded the Paleontological Society Medal, the society's highest honor.

Archives of the Paleontological Research Institution

Tilly Edinger (1897-1967) was born in Frankfurt, Germany. Her father was a neurologist, so she grew up surrounded by science and scholarship. She earned a Ph.D. from Frankfurt University in 1921 and then found unpaid work in German museums. Edinger was a Jew and was forced to flee Nazi Germany in 1939. In 1940, she was offered a position at Harvard University's Museum of Comparative Zoology, where she spent the remainder of her career.

From the Archives of the Museum of Comparative Zoology, Ernst Mayr Library, Harvard University

Mary Dawson (alive and digging!) studied at Michigan State University for her undergraduate degree and at the University of Kansas for her Ph.D. She has spent most of her career as a curator at the Carnegie Museum of Natural History, focusing her research on ancient mammals. Dawson is a recipient of the A.S. Romer-G.G. Simpson Medal, the Society of Vertebrate Paleontology's highest award. She is currently Curator Emeritus of Vertebrate Paleontology at the Carnegie Museum.

Copyright Carnegie Museum of Natural History

Karen Chin (alive and digging!) earned her Ph.D. in geological sciences from the University of California at Santa Barbara. Her research is focused on reconstructing ancient ecosystems through the lens of trophic interactions, especially those revealed by analysis of coprolites. She is currently an Associate Professor in the Department of Geological Sciences at the University of Colorado Boulder, as well as Curator of Paleontology at the University of Colorado Museum of Natural History.

Courtesy Karen Chin

Helen James (alive and digging!) completed her undergraduate work at the University of Arkansas and earned her Ph.D. at Oxford University. Her research is focused on the ecology and evolution of birds, as well as their responses on islands to historic human settlement. James is currently Curator in Charge in the Division of Birds at the Smithsonian Institution National Museum of Natural History's Department of Vertebrate Zoology.

Helen James, Research Zoologist and Curator of Birds, National Museum of Natural History, Smithsonian Institution Photo by James Di Loreto

Phoebe Cohen (alive and digging!) studied Earth System Science as an undergraduate at Cornell University. She received her Ph.D. from Harvard University, where she began researching the fossil record of Proterozoic eukaryotes. Cohen is a recipient of the Geological Society of America Subaru Outstanding Woman in Science Award. She is currently an Assistant Professor of Geosciences at Williams College.

Courtesy Phoebe Cohen

Glossary

astrobiology: the scientific search for life beyond Earth

ballast: material carried by hot air balloons to control how quickly and how high they rise into the air

biomineralization: the process by which living organisms produce internal or external skeletons, such as shells or bones

bryozoans: small aquatic animals, living in groups, that use tentacles to grab food particles from the water

carnivore: an animal, such as a shark or tiger, that gains its energy by eating other animals

chemical makeup: the elements responsible for the composition of a particular structure; for example, the chemical makeup of a snail's shell is calcium carbonate ($CaCO_3$)

complex life: organisms that have a cell nucleus that houses genetic information (eukaryotes); generally, organisms other than bacteria and viruses are considered to represent complex life

corals: ocean animals living alone or in large groups that grab small bits of food from the water with stinging tentacles; some form gigantic reef ecosystems

crinoids: ocean animals, sometimes called sea lilies, that look like plants; they use long, feather-like arms to capture their food out of the water

crust: the uppermost layer of the earth, including the continents and ocean floor

Crustacea: a group of invertebrate animals that includes crabs, lobsters, shrimp, and pill bugs

coprolites: fossilized poop

curator: a person whose job is to organize and care for museum collections

echinoderms: a group of invertebrate ocean animals that includes starfish, sea urchins, crinoids, and sea cucumbers, as well as a variety of extinct forms

electron microscope: aa powerful type of microscope that allows viewing of things that are too small to be seen with a regular microscope

endocast: a natural or human-made copy of a soft-tissue organ that once filled a space inside an organism; for example, a replica brain made by pouring rubber into the hollow space inside a fossilized skull

entomologist: a scientist who studies insects or other bugs

eukaryote: single- or multicellular organisms with complex cells that have a cell nucleus

evolution: the scientific theory that organisms have changed over time and are descended from shared common ancestors that lived in the past (descent with modification)

feces: poop, or dung

Foraminifera: very small, single-celled organisms related to amoeba that often produce external shells

herbivore: an animal, such as a deer or rabbit, that gains its energy by eating plants

introduced: an organism that did not originate in a region, but instead was brought there by people

invertebrate: an animal that does not have bones; examples include insects, snails, worms, starfish, jellyfish, and corals

lava tube caves: a hole sometimes left in the earth by lava flows associated with volcanic activity

mantle: the internal layer of the Earth below the crust; the flow of molten rock within the mantle moves the rigid crust above, driving plate tectonics

microfossils: fossils that are too small to carefully observe without the aid of a microscope

micropaleontology: the scientific study of very small fossils, for example, Foraminifera

mollusks: a diverse group of invertebrate animals that includes snails, slugs, clams, octopuses, and squids, along with other modern and extinct forms

multicellular organism: an organism with a body made up of many cells

native: an organism that that originated in the region where it is found

organisms: individual lifeforms

paleoneurology: the scientific study of the nervous systems and brains of extinct animals

petrification: a type of fossilization that causes some parts of ancient organisms to be turned into stone; petrified wood is an example

petroleum: a geologically formed fossil fuel derived from the remains of ancient organisms such as algae; petroleum is used to produce gasoline

plate tectonics: the scientific theory that Earth's surface is divided into a series of plates that move relative to one another, resulting in the formation of mountain belts and volcanoes, and causing earthquakes

prehistoric: the long span of time before modern human civilization

single-celled organism: an organism with a body composed of one cell

species: a genetically similar group of organisms that are able to reproduce with one another

specimens: remains of modern or fossil organisms that are stored in museum collections

stratigraphic correlation: the scientific process of matching rock layers of the same age over long distances; this is often done by comparing the kinds of fossils found in the rock layers

stratigraphic sequence: the layers of rock present at a particular location that may show change over time, indicating evolving ancient habitats

undescribed: a modern or fossil species that scientists have not yet written about or given a formal scientific name

vertebrates: animals with bones; examples include fish, frogs, alligators, birds, elephants, and humans

Beth Stricker has been immersed in the history of paleontology since 2013, when she was first inspired by the adventures of Carlotta Maury. As Director of Exhibitions at the Paleontological Research Institution, she founded the Daring to Dig project, which included exhibition design plans, a website (daringtodig.org), interviews with living paleontologists for a historic archive, and this book. Beth is currently a Project Director at the Smithsonian Institution Traveling Exhibition Service. She has a BA in Classical Civilizations from Hobart and William Smith Colleges, a MFA in Museum Exhibition Planning and Design from the University of the Arts, and has received awards for Innovation in Interpretation from the Museum Association of New York for her playful exhibitions *Did Dinosaurs Poop?* and *Quirks of Nature*.

Alana McGillis is a doodler from Cape Cod, MA, where there are a lot of shipwrecks and horseshoe crabs. She studied Geology at Smith College before working at the Paleontological Research Institution, where she spent more time doodling trilobites than doing her real job. This is her first book! Her favorite paleontologists in it are Tilly Edinger (because she was brave) and Winifred Goldring (because she made pants for her friends). She is available for hire as a fossil hunter, baker, dog walker, and illustrator. More of her doodles and science illustrations can be found at www.alanamcgillis.com.